Diomedes Ángel Castro Laura

SOLAR IRRIGATION SYSTEM EFFICIENCY

Diomedes Ángel Castro Laura

SOLAR IRRIGATION SYSTEM EFFICIENCY

at different depths in the cultivation of broad bean (Vicia faba L.) in the province of ACOBAMBA

ScienciaScripts

Imprint

Any brand names and product names mentioned in this book are subject to trademark, brand or patent protection and are trademarks or registered trademarks of their respective holders. The use of brand names, product names, common names, trade names, product descriptions etc. even without a particular marking in this work is in no way to be construed to mean that such names may be regarded as unrestricted in respect of trademark and brand protection legislation and could thus be used by anyone.

Cover image: www.ingimage.com

This book is a translation from the original published under ISBN 978-620-2-10330-5.

Publisher:
Sciencia Scripts
is a trademark of
Dodo Books Indian Ocean Ltd. and OmniScriptum S.R.L publishing group

120 High Road, East Finchley, London, N2 9ED, United Kingdom
Str. Armeneasca 28/1, office 1, Chisinau MD-2012, Republic of Moldova, Europe
Printed at: see last page
ISBN: 978-620-5-55916-1

TITLE

"EFFICIENCY OF THE IRRIGATION SYSTEM BY SOLARIZATION AT DIFFERENT DEPTHS IN THE CULTIVATION OF BROAD BEAN (VICIA FABA L) IN THE PROVINCE OF ACOBAMBA".

AUTHOR

Bach. LAURA CASTRO, Diomedes Angel.

CONSULTANT

Mtro. Jesús Antonio, JAIME PIÑAS

DEDICATION

▶ To the divine Lord for giving me life and guiding me on the right path during my professional training.
▶ To my parents: Darío and Marina for the support economic y many lessons

▶ Also to the training house of
studies at the National University of Huancavelica, which allowed me a correct formation during my studies.
permanence.

ACKNOWLEDGMENT

A very special thanks to my parents: Dario and Marina, for their infinite unconditional economic support and to my brothers Heber and Anderson for their persevering support in the most difficult moments of economic and moral sharing.

To. JAIME PIÑAS JESÚS ANTONIO, with appreciation and affection as my advisor, who with his valuable knowledge and experience guided me in the execution and writing of the final report of the research project. Finally, to our house of studies to the National University of Huancavelica, to the teachers for their shared knowledge and experience of the professional school of Agronomy and to the friends for having shared with their admiration of their valuable knowledge for my professional training.

RECOGNITION

To the National University of Huancavelica, Faculty of Agrarian Sciences, Professional School of Agronomy, for giving me the opportunity to be a student of this school of Agronomy that has been imparting knowledge and knowledge of this noble profession with quality to all its students.

INDEX

SUMMARY

Objective: To know the efficiency of the irrigation system by solarization at different depths, in the cultivation of broad bean, in the province of Acobamba **Method**: The scientific research method was based on work and evaluation of irrigation by solarization of broad bean by means of PET bottles installed at different depths of cover and cabinet, evaluating, tabulating and inferring the results. The present work of exploration determined an **explanatory** level of causal relationship since it sought to evaluate the yield of the bean crop relating its response to irrigation by solarization at three levels of depth of the PET bottles.

Results: It can be deduced that the production of green beans on average using three seeds per stroke ensures a yield of **9.25 t/ha, even with a 2% loss due to bird damage.** Even with a 2% decrease due to bird damage, this result of the solarization irrigation system offers advantages to the producer who cultivates under rainfed or rainfed conditions, being the water demand / hectare of the system with cover at 0.06 m., at ground level of 11.81 m3 /ha., at 0.12 m., 7.13 m3/ha., and 4.86 m3/ha., at 0.18 m., during the entire crop cycle.

Conclusions: It is inferred that the solarization irrigation system allows irrigation in an efficient way to the crop and implies a water saving greater than **100%** with respect to the exclusive use of traditional irrigation systems whose demand throughout the crop cycle allows us to infer an average consumption of **7.93 m3/ha,** We can also conclude that irrigation by solarization is very efficient, simple and economical, characterizing an optimal irrigation technology for crop production under the rainfed production regime, **leading to an increase in production and reducing the demand for irrigation water** to a minimum level of requirement per hectare with 95% and 99% probability.

Key words: Evaluation, solar irrigation, PET bottles.

INTRODUCTION

Water for irrigation is a resource that every day is limited both quantitatively and qualitatively due to the accelerated growth of the demands for family, livestock, agricultural and industrial use that is still incipient in Acobamba Huancavelica, in recent years there is a trend of maximizing the efficiency of water use for irrigation using migratory technology such as sprinkling, drip, micro sprinkling, sweating, xylem irrigation, among others.

However, these technologies are not adaptable to the configuration of the terrestrial relief of Acobamba and even less to the micro-parceling of the agricultural properties of small farmers, making this attempt useless, so without a change in traditional or surface irrigation technology, what we need is a more rational use of water, which should be used more efficiently in irrigation systems.

Irrigation efficiency is that capable of providing the soil with permanent and timely moisture within the appropriate water needs and without reaching a stress due to lack of water that harms the crop, this will depend on the typologies of the crops in process, climatic conditions, management and development environment.

All of which can be expressed through the **efficiency of irrigation water application, which is the amount of useful water for the crop that remains in the soil after irrigation**, in relation to the total water that was applied, the efficiency of application in the plot is generally measured as a percentage or liters of useful water in the soil for every 100 liters applied. Solar irrigation allows maintaining constant humidity in areas surrounding the root system of the crop, **which will be sufficient for food production without causing water deficit,** is one of the questions that should be clarified with this research.

Also, taking into account that only a percentage of less than 20% of the world's agricultural land is under irrigation and, being these the ones that provide 40% of the world's agricultural production, it is essential to generate new technologies that increase the efficiency of irrigation water application to the crop, depending on climatic factors, crop phenology and soil characteristics.

Agriculture is the production system with the highest global water demand; being irrigation the activity that consumes 70% of this resource worldwide, it is essential to improve irrigation efficiency, supplying the plantation with only the amount of water necessary to satisfy its needs without causing water loss by deep percolation with the consequent washing of fertilizers.

In view of the above, it is necessary to carry out studies of **"Efficiency of the irrigation system by solarization at different depths in the cultivation of broad bean (Vicia faba L.) in the province of Acobamba",** due to the simplicity of the process and indirect promotion of environmental conservation.

CHAPTER I

PROBLEM STATEMENT

1.1. Description of Problem

Irrigation water management requires adequate estimates of water requirements, which means knowing precisely the water needs of crops at different stages of their development as a crop in process, as well as new technologies if we intend to have satisfactory yields despite the low availability of water resources in Acobamba. The introduction of new irrigation technologies has been scarce and insufficient to cover the prevailing climate and soil conditions in the area. Fava bean (Vicia faba L.) is an economically important crop for Acobamba farmers and under local conditions indicates a crop that is well adapted to environments of low water availability and intense temperature variations. However, the water needs of this crop have not been covered due to the deficient availability of this water resource. Another important aspect to consider is the increasing contamination of the Acobamba district due to poor solid waste management, population growth, and the existing informal agricultural trade.

1.2. Background of the problem. -

In 1991, the New Irrigation Project of the Studies Directorate of the National Small Irrigation Program (PRONAPEMI) prepared the feasibility study with designs at the construction level for the Acobamba I Stage Irrigation Project. By means of the Study Review Act of May 30, 1991, the Elaborating Commission designated with Memorandum N° 06-91-AG-91- AG-DGAS PRONAPEMI-DEST/PRO, recommends the approval of the Feasibility Study with Designs at Constructive Level of the Acobamba I Stage Irrigation Project. With official communication No. 311-91 AG-DGAS-PRONAPEMI-DE, dated 20 June 1991, the Executive Direction of the National Program of Small and Medium Irrigation PRONAPEMI requests the approval of the Feasibility Study with design at constructive level of the Acobamba Irrigation Project I Stage. Geological Study - Final Report Tunnels 1, 2, 3 and 4 of the feasibility study with designs at constructive level of the Acobamba Irrigation was carried out by PRONAMACHCS in 1995. Feasibility Study General Fund Contravalor Peru Canada - NGO PRODER 1997. On July 9, 2000, the National Public Investment System was created by Law No. 27293. Between 2003 and 2004 the Huancavelica Regional Government carried out excavation works for Tunnel No. 04 from progressive 40+406.59 to 41+ 480.59 making a total of 1,074.0m of progress. The updated feasibility study of the Acobamba Irrigation Construction Project was sent by the Huancavelica Regional Government with official letter N° 122- 2004/GOB. REG. -HVCA-GRDE. With Official Communication N° 3249-2004-AG-OGPA-OI of November 04, 2004, Technical Report N° 161-2004-AG-OGPA/OI was sent to the Huancavelica Regional Government, containing the observations to the Updated Feasibility Study of the Acobamba Irrigation Construction Project. The Formulating Unit is the Huancavelica Regional Government. The Acobamba Irrigation Completion Project Profile was sent by the Huancavelica Regional Government with official letter N° 538-2006 / GOB. REG. HVCA / GRPP and AT-SGPI and CTI. With Official Communication N° 3620-2006-AG-OGPA-OI, Technical Report N° 256- 20056-AG-OGPA-OI was sent, containing the observations to the profile of the Acobamba Irrigation Termination Project. With Official Communication No. 711- 2006/GOB.REG.HVCA/GRPP and AT-SGPI and CTI, the Regional Government of Huancavelica sent the Project profile with the observations' acquittals. In order to be compatible with the approval of the profile and to promote a lower investment with the real benefits of the proposal, it was agreed to prepare a Technical Report to support the feasibility of incorporating a new alternative, which would allow the execution of a pre-feasibility study instead of directly preparing a feasibility study. This was agreed in a meeting held on July 13, 2007, with the participation of the Mayor of the Provincial Municipality of Acobamba, the Infrastructure Manager of the Government of Huancavelica and the Infrastructure Manager of the Regional Government of Huancavelica. Regional Huancavelica, Sub Regional Manager of Acobamba, Regional Councilor, OPI Huancavelica Representative and DGPM representatives **(Jaime et. al (2008) pp. 9).Scope of the Acobamba Irrigation Project** The Acobamba Irrigation Project was included in the 2013-2015 Provincial Strategic Plan and included the collection and direct use of water from the Dos de Mayyo, Tinquer and Paucará rivers; it also includes the diversion of the Huarmilla river to the Dos de Mayo river basin, supported by discharges from the Alpachaca, Huarmilla, Escalón and Huaripaccha springs during low water levels. All these events made the province of Acobamba Huancavelica in 2007 experienced an unusual interest in the possibility of having the water resource for irrigation of their crop areas, since with a budget of S/. 20, 000,000.00 (twenty million nuevos soles) allocated by the Central Government, the farmers were hoping to have enough water resources to irrigate 4,474 hectares of agricultural land under the Acobamba Irrigation Project, which included the construction of four tunnels with a length of 5.53 km and canals with a length greater than 5.53 km, and canals over a length of more than 30 km **(Jaime et. al. (2008) pp. 10).**

Research Project "Evaluation of water resources to derive to the Acobamba Irrigation System", in this situation the Professional School of Agronomy in alliance with the Federation of Agricultural Associations of Acobamba - Huancavelica "FAAPA - Huancavelica". In its eagerness to be an active part of this process of expansion of the agricultural frontier, as part of a research work, an inventory of water resources to be derived to this irrigation system was carried out, since by references it was known that the water supply was deficient to meet the critical demand of the project, this evaluation of these resources was carried out in two well-defined periods: low water season and the end of the rainy season. The research lasted 12 months (March 2007 - February 2008) and it was determined that the maximum volume of water that could be derived in the low water season from June to September only reached 285.52 liters/second, which would only ensure the irrigation of 125.23 hectares of crops in the small season, discarding as such all the water that could be used in the low water season. possibility of irrigating larger areas if we only take as an alternative to derive water from the collection and direct use of water from the Dos de Mayo, Tinquer and Paucará rivers; in addition to the Huarmilla river diversion to the Dos de Mayo river basin, supported in low water with discharges from the Alpachaca, Huarmilla and Escalón streams; and the Huaripaccha spring. **(Jaime et. al (2008) pp. 7).**

Conclusions, a deficit of water resources was inferred to incorporate **1,300 ha** in small season to the irrigation system, since the demand calculation for the critical season at that time only represented an irrigation module of **2.3 l/s./ha.,** which considering the volume of water available **285.52 l/s** would only allow the irrigation of **124.14 ha.** in small season. In addition, it was stated that the water resource available in the critical season, determined at **285.52 l/s,** was insufficient to incorporate 1,300 ha to the Acobamba irrigation system even using advanced irrigation technology such as irrigation by injection to the Xylem 99% efficiency (chemotherapy or ultramicro), hydrosorb (US $ 0.10/plant), Sweating, Microaspersion, Pulse Jets, Sprinkling and Drip **(Jaime et. al (2008) pp. 30).** These antecedents were crucial to carry out the present research as a possibility of irrigated production using the solarization irrigation method.

1.3. Problem formulation

How efficient will the solar irrigation system be in the cultivation of fava beans in the Province of Acobamba - Huancavelica?

1.4. Objectives:

1.4.1. General.

► To know the efficiency of the irrigation system by solarization at different depths of cover, in the cultivation of broad bean, in the province of Acobamba.

1.4.2. Specific:

✓ Evaluate water demand in the solar irrigation system in the province of Acobamba.

✓ To evaluate the bean yield obtained with the solarization irrigation system in the province of Acobamba.

✓ Recycle disposable beverage containers for use in irrigation in the province of Acobamba .

1.5. Justification

Scientific:

The research aims to verify the direct problem referred to the water consumption of bean crops given the scarce condition of this resource for irrigation in the province of Acobamba, since the allocation provided by the rains is mostly insufficient to cover the needs of the crop in the plot. The present exploration work seeks to introduce a simple technology to solve these problems of water shortage for irrigation. In addition, it constitutes an important source of scientific knowledge that should be encouraged, validated and systematized for its dissemination in the correct approach of application by the producers of the area.

Social:

The result of the research will provide stability, individual and family wellbeing, greater social and community participation of producer organizations in rural communities, reviving ancestral values, and finally, it will promote farmer training and creativity in agricultural activities and especially low-cost solar irrigation technology.

Environmental:

A significant fraction of the freshwater consumed by mankind is used for crop irrigation, and much of it is wasted in inefficient irrigation systems. Although there are efficient irrigation systems, such as conventional drip systems, they tend to be expensive and are therefore within the reach of few farmers, especially in less economically favored countries. The solar drip system changes this situation by offering a very efficient, high quality, simple and inexpensive to install irrigation system that can help millions of farmers around the world to increase their production using much less water, recycling plastic containers, available on site.

Economic:

It has an impact on the economic valorization of the water resource, constitutes the basis for its rational use and crop planning, and increases the profitability and quality of agricultural activity. In general, research is very important to improve and conserve water quality, in order to achieve the maximum economic and sustainable benefit; consequently, to avoid its degradation, which has an unfavorable impact on ecological and social stability.

1.6.- Limitations. -

Acobamba is geographically located in the central highlands of Peru being located with respect to the GRENWICH meridian at: West Longitude: 74° 31'48 and South Latitude: 12° 54'17 with respect to the equator. The Provincial surface of Acobamba covers 910.82 **Km2**, which represents 4.03% of the total extension of the Department and Region of Huancavelica which is **22,557.47KM**. In general, the province is located at an altitude of 2,200 meters above sea level on the banks of the Mantaro River on the east side and 2,800 meters above sea level on the west side. m.a.s.l. on the banks of the Santa Ola River; the highest part of the province is in the District of Paucará at more than 4,000 m.a.s.l. **The capital of the province Acobamba is located at 3,423 meters above sea level.** Its areas of agricultural aptitude have favorable soil and ecological conditions, an average annual rainfall of 700 mm, an average annual temperature of 12 ºC and an average relative humidity of 60%. However, one of the critical problems in Acobamba is the lack of irrigation water that does not allow planning a double agricultural production campaign and increasing the rate of use of its available crop areas.

CHAPTER II

THEORETICAL FRAMEWORK

2.1.- Background

The technique corresponds to a modern adaptation of ancient irrigation methods that buried earthenware containers near the plant. This system is a little more laborious and the roots can reach the reservoir, which can be an advantage if the water is good. With this practice adopted and called solarization irrigation system, outdoors it favors the production of distilled water which is enriched by solar radiation, similar to rain when its water drops fall to the ground; we can avoid salts, nitrates and other pollutants, underground or above ground, in both cases we not only save a lot of water, but we can also save time and money. It should be noted that in the area of Huancavelica and the Province of Acobamba, there have been no studies of irrigation by solarization by institutions such as the Ministry of Agriculture, universities, producers, etc..

2.2. Theoretical bases

2.2.1. General

The solarization irrigation system is an ingenious way, which produces distilled water by evaporating the water contained in a vessel inside the empty bottle of 01 gallon capacity by the effect of solar radiation. For irrigation by solarization, recycled glass or plastic bottles or containers can be used. This traps the evaporated water from the soil and from the reservoir inside. The humidity condenses on the inner walls of the bottle and falls or slides in the form of drops and is directed by its walls back to the ground. At night it has a double function as it also collects a certain amount of dew.

This irrigation proposal makes it possible to save enormous amounts of water for irrigation and also allows us to cultivate plants that need timely and better quality water for their growth. The water produced by the system does not contain salt, nitrates or any other pollutants harmful to plant cultivation. **FLORES (2016), describes in his thesis** "Estudio del uso de botellas plasticas recicladas **(Politereftalato de etileno) (PET)** en el riego por goteo solar y su aplicaciòn en la forestaciòn", Bolivia, 7 pp., The Kyoto Protocol on climate change, which is a protocol of the United Nations Framework Convention on Climate Change (UNFCCC) and an international agreement that aims to reduce emissions of six greenhouse gases that cause global warming: carbon dioxide (CO2), methane gas (CH4) and nitrous oxide (N2O), and the other three are fluorinated industrial gases: hydrofluorocarbons (HFCs), perfluorocarbons (PFCs) and sulfur hexafluoride (SF6), where it states in its Article 2:, (iv) research, promotion, development and increased use of new and renewable forms of energy, carbon dioxide sequestration technologies and advanced and novel technologies that are environmentally sound. (United Nations, 1998)

2.2.2. Concepts

2.2.2.1.-Cycle of water

PADILLA (2020) mentions that the water cycle, also known as the "hydrological cycle, is the process of transformation and circulation of water on Earth". In this sense, the water cycle consists of the transfer of water from one place to another, changing its physical state: from liquid to gaseous or solid state, or from gaseous to liquid state, depending on environmental conditions, he also **refers** that on Earth, water is distributed in the seas, rivers or lakes in liquid state; in the glaciers of the poles and mountains in solid state, and in the clouds, in gaseous state. Water evaporates from the seas due to the heat of the sun, leaving behind its salts. Winds carry the water vapors to the ground, where they gather in clouds that condense as they cool and fall as rain, making it possible for plants and, consequently, animals to grow. Once the water reaches the ground, by one route or another it returns to the sea, where it mixes again with the salt water. We humans as well as many other land animals, We interrupt that cycle, we intercept the water somewhere in its passage between the sky and the sea. We use it and almost invariably, we pollute it, and then allow it to continue on its way to the ocean. As the years go by, the amount of water we take out of its natural cycle to use it is increasing. The flows are now enormous.

2.2.2.2.2.-Watering

FLORES (2016), in his thesis "Estudio del uso de botellas plasticas recicladas **(Politereftalato de etileno) (PET)** en el riego por goteo solar y su aplicaciòn en la forestaciòn", Bolivia, 8 pp., **refers** that irrigation consists of the artificial application of water for plants to grow and develop. Basically, it allows controlling the effects of frost and drought, as well as optimizing the use of water (Intermediate Technology Development Group, ITDG, 2011), **adding** that irrigation is the timely supply of the right amount of water to crops so that they do not suffer a decrease in their yields and without causing damage to the environment (Chipana Rivera, 2003).

JAIME (2017) pp. 102, mentions that irrigation is required when the amount of rainfall is insufficient to compensate for water losses due to evapotranspiration. The main objective of irrigation is the application of water at the required time and with the precise amount of water. By calculating the daily water balance in the root zone of the soil, irrigation rates and times can be planned. To avoid water stress, irrigation should be applied before or at the moment of exhaustion of the easily extractable soil water sheet (Dr,i,Afa). On the other hand, to avoid percolation losses that may result in the flushing of important nutrients from the root zone, the net irrigation lamina should be less than or equal to the depletion of moisture in the root zone of the soil (Ii,Dr,i). (FAO Irrigation and Drainage No. 24 1990 p. 171).

2.2.2.3.-Importance of irrigation

JAIME (2014) pp. 9, states that the importance of crop irrigation is evidenced by the fact that it contributes one third of the gross value of agricultural production, so the need to apply sustainability criteria in the use of water obliges the agricultural sector to adjust and rationalize its water consumption in irrigation systems through an adequate calculation of crop water demand for an adequate irrigation scheduling in order to allow a better use of this resource, maintain the soil with sufficient moisture for the correct development of the crop, as well as avoid water losses through surface runoff and deep percolation, a situation that in addition to reducing the inadequate use of the resource, mitigating the effects of drought, will reduce the problems of contamination and overexploitation.

2.2.2.4.-Irrigation by gravity

As FLORES (2016, p. 9) indicates, gravity irrigation is characterized by the application of a sheet of water that moves by gravity and slides on the ground following the slope, without requiring extra energy to move. In this method of irrigation, water flows through large canals to the distribution centers that distribute it to the plots through medium and small ditches; the water is distributed and directed using slats, stones with mud or sluice gates.

2.2.2.5.-Irrigation by aspersión

According to FLORES (2016, p. 9) sprinkler irrigation is a method that applies water in the form of spraying simulating rain, in a controlled manner in time and intensity, adding that sprinkler irrigation, in addition to allowing management of the amount of water to be used, helps prevent frost and mitigate the effects of droughts. It also allows adding fertilizers or phytosanitary products to the water (Intermediate Technology Development Group, ITDG, 2011).

2.2.2.6.-Localized Irrigation

FLORES (2016, p. 9) refers, that localized irrigation is the application of water to the soil, in a more or less restricted area of the root volume, he also adds that according to (Rodrigo Lopez & others, 1997 cited in his thesis) it is characterized by:
1) It does not wet, in general, the entire soil by applying water on or under its surface.
2) It uses small flow rates at low pressure.

3) Apply water near the plant through varying numbers of emission points.
4) High frequency irrigation in order to maintain soil moisture.

FLORES (2016, p. 9) describes, that localized irrigation, is a type of drip irrigation with a flow rate no greater than 16l/h per point of emission or linear meter of drip hose cited in his thesis (Rodrigo López & others, 1997). Drip irrigation provides localized and constant water drops to plant roots through emitters commonly known as "drippers". Its discharge fluctuates in the range of 2 to 4 liters per hour per dripper. The application of small quantities of water allows maintaining a stable level of humidity, thus minimizing water consumption and losses due to evaporation and filtration. Drip irrigation guarantees at least 60% water savings, information cited in his thesis "Study of the use of recycled plastic bottles (PET) in solar drip irrigation and its application in forestry" (Intermediate Technology Development Group, ITDG, 2011).

2.2.2.7.-Drip Irrigation System

It is a localized irrigation method where water is applied in the form of drops through emitters, commonly called "drippers". Discharge from the emitters fluctuates in the range of 2 to 4 liters per hour per dripper. Drip irrigation delivers small amounts of moisture at frequent intervals to the root of each plant through thin plastic tubes.

2.2.2.8.-Solar irrigation by goteo

PADILLA (2020) reports that Solar Drip Irrigation is a technique by which PET plastic bottles can be reused to water plants without losing soil moisture. Two bottles of different sizes, a liter and a half of water and a cutter are used. The bottles are cut and the smaller one is placed in a hole in the ground next to the plant with a liter and a half of water, then the large bottle with its cap is placed on top of the small bottle containing the water, finally the hole is filled with the same soil. Thanks to this simple device, the soil can be kept moist for a period of approximately 45 days without having to water constantly. It works thanks to the heat of the sun that evaporates and condenses the water inside the large bottle and the liquid drains into the soil without being lost in the atmosphere.

PADILLA (2020) mentions that solar irrigation is a domestic irrigation technique that tries to take advantage of the evaporation that occurs in an enclosure closed by a transparent plastic cover, inside which there is a rise in temperature due to the greenhouse effect, to irrigate through evaporation and condensation that occurs inside the plastic cover. In relation to the operation **PADILLA (2020) describes** that the operation is simple, the system consists of the following parts: transparent plastic cover, container for the irrigation water; in both cases PET bottles of different sizes. The water used for irrigation is poured into the container installed for this purpose, and once the plastic cover is in place, the water evaporates slowly during the day, causing the relative humidity inside the cover to rise. When the humidity comes into contact with the roof wall it condenses, because the wall is colder than the air inside the roof. This condensation produces a continuous drip along the wall of the roof, causing continuous watering on the surface inside the **roof. He** also **states** that condensation occurs at night, when the external temperature drops and the roof cools down, causing condensation on the roof wall. condensation occurs because the wall is colder than the air inside the roof. This condensation produces a continuous drip along the roof wall, causing continuous watering on the surface inside the roof. Condensation occurs at night, when the outside temperature drops and the roof cools, causing condensation on the roof wall.

2.2.2.9.-Solarization or Drip Irrigation System Solar

PADILLA (2020) narrates that the irrigation system by solarization or solar drip also known as Kondenskompressor or komkom is an irrigation technique designed to achieve an optimal use of water by using the sun's energy as a driving force in the process of distillation and movement of water. It is a system of surprising simplicity and efficiency by means of which it is possible to reduce the amount of irrigation water by up to 10 times compared to traditional irrigation systems.

This system also has the advantage of making it possible to use saline water or even seawater for irrigation, since it transforms salt water into fresh water.

2.2.2.10.-Operation of the Solar Drip Irrigation System PADILLA (2020) refers that the operation of the Kondenskompressor is similar to that of the

solar stills. When the sun's rays hit the Kondenskompressor, a greenhouse effect is triggered inside the Kondenskompressor, and as the temperature rises, the water in the tank evaporates. The water condenses on the walls of the carafe. When the Kondenskompressor is in the sun, the water is constantly evaporating, so the drops get bigger and bigger, until they start to fall down the walls of the carafe to the ground, wetting it completely.

2.2.3.- Cultivation of fava bean

2.2.3.1.-Origin and History

Cerrate et al, (1981), report that fava bean cultivation is considered to be native to Central Asia, the Mediterranean and Abyssinia. Many consider it to be native to Europe, where it has been cultivated since pre-historic times, and as a secondary center, North Africa, where it is widely distributed. Christopher Columbus on his second voyage brought this crop to America, where it was planted in the Antilles, but did not prosper due to environmental differences in its place of origin. With the conquest, the Spaniards introduced it in the Peruvian highlands, where, due to favorable conditions, the crop was able to spread.At the beginning, a great variability of forms was cultivated, which were selected in a natural way, discarding those that were not adapted to the environment. New types and forms of beans appeared, different from the original ones, which constitute a valuable source of possible genes for selection **(Marmolejo and Suasnabar, 2000).**

2.2.3.2.2.-Taxonomy of the Plant:

Aliaga (2004), makes a taxonomic classification as follows:

DIVISION Spermathophyta (Magnoliophyta), (Phanerogams).

SUB DIVISION : Angiospermae (Magnoliophytina).

CLASS: Dicotyledoneae (Magnoliopsida).

SUB CLASS: Archichlamideae or (Coripetalas).

ORDER: Fabales (Leguminosales).

FAMILY: Fabaceae (Leguminoceae).

GENUS: Vicia.

SPECIES: Vicia faba L.

2.2.3.3.-Plant Morphology:

The broad bean is an annual plant with a more or less erect habit, which is described below:

Root System:

Camarena et al. (2003) stipulate that the main root is a tap root, which manages to go deep into the soil relatively quickly. The root system is definitely quite vigorous, generating long lateral roots from the tap root; it can reach up to 1m deep, but its growth is usually in the first 50 to 60 cm of soil. **Likewise, Marmolejo and Suasnabar (2000) mention**: In the secondary roots, nodules are usually formed, where atmospheric nitrogen-fixing bacteria are housed.

The Main Stem and Branches:

Camarena et al. (2003) report that the stems are erect, robust, hollow and can reach up to 2 m in height, although the normal height is between 0.8 and 1.20 m. From the basal nodes of the main stem may originate between 01 to 05 branches per plant; the average number depends mainly on the population density, soil fertility and planting date, but in general is close to 03. Most of the branches begin to develop early after emergence, becoming visible when the main stem has an average of 03 leaves. The basal branches, which are generally quite vigorous, attain a growth that in many cases resembles that of the main stem; the basal branches contribute, on average, between 50 and 70% of the total pods produced by a plant. As the branches become heavier as the grain fills, they move away from the central axis and may even break. This situation hinders, to a greater or lesser extent, the work of harvesting the pods.

Marmolejo and Suasnabar (2000), state that the stems are erect, fistulous and robust.

Vegetative and reproductive nodes:

Camarena et al. (2003), state that the number of vegetative nodes is normally 05 to 07. The average number of vegetative nodes on basal branches, meanwhile, generally varies between 03 and 04. Reproductive nodes are produced in large numbers, with little difference between the amount produced by the main stem and by each basal branch; the average number of reproductive nodes per stem varies between 12 and 18.

The Leaves:

Camarena et al. (2003) state that the leaves are compound, more or less oval and grayish-green in color; they are arranged alternately and do not have tendrils. The basal leaves begin to die during the grain filling stage and this situation continues to occur gradually in an ascending manner; death occurs mainly due to the shading of the leaves and the end of root activity, increases in temperature and the presence of Botrytis fabae.

Marmolejo and Suasnabar (2000), indicate that they are pinnate composites with 4 to 7 glabrous leaflets of entire edge which are almost always wide, more or less oval, grayish green; semi-sagittate stipules. The rachis is well developed and is considered the median axis of the leaf; the leaflets are almost directly inserted due to the lack of petiole. The leaf is attached to the stem through the petiole at the stem node. The petiole is well differentiated by its elongated shape. The stipules are appendages that arise at the base of the leaf, they are semi-sagittate and their purpose is to protect the buds.

Flowers and Inflorescences:

Camarena et al, (2003), describe that bean flowers are large and are arranged in inflorescences that correspond to short bunches. In a plant the number of flowers per bunch reaches an average that varies between 03 and 04, **they also confirm** that, although most of the bunches produce between 03 and 05 flowers, it is common to obtain between 00 and 02 pods per node. In this sense, in 80% or more of the reproductive nodes there is a total fall (abscission), either of flowers or young pods. This situation determines that the average number of pods per node, at the level of the whole plant, is very low. According to **Camarena et al. (2003),** although each plant can produce more than 300 flowers, considering a total of 80 to 100 total reproductive nodes per plant, the fruit set percentage does not exceed an average of 10 to 15%.

On the other hand, **Camarena et al, (2003),** due to the type of flowering, **indicate** that due to the extensive vegetative period of the bean plant, there comes a stage in which the plants present simultaneous growth of stems and leaves, opening of flowers, growth of pods and filling of grains.

Camarena et al, (2003), with respect to pollination in the broad bean plant, **state** that this occurs cross-pollination, and can even reach values as high as 70%. The percentage will depend on the cultivar, climatic conditions and the population of pollinating insects. In any case, cross-pollination generally reaches between 30 and 50 %.

Marmolejo and Suasnabar (2000), in turn, **reveal** that the inflorescences are of racemose type of axial origin. It originates in a short developed peduncle, followed by the rachis where the flowers are inserted by means of the pedicels, which are pedunculi that support the flower, which are very small. The flowers are inserted and hang from only one side of the rachis.

The flower, originating in the axils, grouped in short clusters, in number of 2 to 12 flowers with a particular color, although not intense. Corolla dialipetalous, with 5 unequal petals, white or bluish, with black or brownish spots on the two wings and characteristic brownish stripes on the banner; the keel which are 2 slightly colored, pale green glabrous calyx, tube-shaped and with 5 teeth. There are 10 stamens, 9 of them solitary and their filaments form a tube enclosing the pistil, the tenth stamen remains free (diadelphous). Cross-pollination in 60 to 30% is carried out by bees, which affects varietal purity.

The Flowering Stage:

According to Camarena et al, (2003), flowering is prolonged for a long period (60 to 75 days in crops sown at optimum dates), the inflorescences of the first reproductive nodes have fewer flowers (03 to 04 per node), being normal, in early sowings when flowering occurs with temperatures still very low, that none of them thrive. From the third reproductive node onwards, the number of flowers increases and fruit set becomes more likely. Summerfield et al. (1991) since this is a species whose flowering starts more quickly on long days, but is not inhibited (qualitative character) under short days, but only delayed.

The Fruit:

Camarena et al, (2003), explain that pods or legumes correspond to fruits, which are straight and fleshy in their initial stages, the length of pods, fluctuates approximately between 12 and 35 cm, the width of the pods, varies between 02 and 2.5 cm on average.

The number of seeds per pod varies considerably depending on the position of the pods on the stems. Thus, those of the lower nodes manage to produce a greater number of seeds than those of the upper nodes, producing between 02 and 05 seeds per pod on average, with an individual maximum of up to 07 seeds in the lower pods and a minimum of two in the pods located in the highest productive nodes. In cultivars producing larger seeds, an average of 03 seeds per pod is obtained.

Marmolejo and Suasnabar (2000) confirm that it is a pod or legume, thick, fleshy, elongated, with seeds arranged in the ventral row. The dehiscence occurs in the dorsal and ventral sutures, separating it into two valves or halves. The pods are green when tender and at maturity become leathery and black. The disposition of the fruit varies, from erect forming a very acute angle with the stem, to hanging. The length varies depending on the varieties and the environment from 5 cm, up to 30 to 40 cm, may contain 2 to 6 compressed or large seeds of different color and size.

The Grain Filling Stage:

Camarena et al, (2003), state that, unlike other grain legumes, the growth of pods and beans occurs simultaneously; the immature beans increase in size until they reach optimum maturity for green consumption with an average moisture content of 72-74%.

The color of the shell is initially green, reaching a color that is usually cream or grayish green when ripe for green consumption, although it can also be relatively bright green. Then the kernels begin to rapidly lose moisture content, becoming increasingly opaque and less greenish in color.

Then **Camarena et al, (2003), concluded** that in cultivars that produce small-sized grains, the weight of 100 grains varies between 70 and 120 gr; in cultivars that produce intermediate-sized grains, the weight of 100 grains for green consumption varies between 160 and 200 gr; finally, in cultivars that produce large-sized grains, the weight of 100 grains at maturity for green consumption is 200 to 250 gr on average.

Seeds:

Camarena et al, (2003), state that, in small-seeded cultivars, 100 dry seeds weigh between 30 and 70 gr. In medium-seeded cultivars, the weight of 100 seeds fluctuates between 70 and 110 g and in large-seeded cultivars, the weight of 100 seeds varies between 120 and 180 gr. The color of dry seeds, meanwhile, can be cream, brown, purple brown, red, black or leaden green.

Marmolejo and Suasnabar (2000), indicate that the seeds are oval in shape, with a smooth, opaque and shiny surface, with a wide variety of colors ranging from dark to light; they can be black, red, green, purple, brownish, grayish, creamy-white or white: they can also be marbled or two-colored, as in the case of a variety "Jilguero de Sicuani". Seed size varies from small, with a length of approximately 1.6 cm in the subspecies, to large seeds with a length of approximately 3.5 cm in the larger subspecies. The weight of 100 seeds varies from 120 to 330 grams, very long hilium of black color generally. Its germination energy is approximately 4 years. The seed is of hypogeous germination. The testa and cotyledons remain underground, this is not frequent since the cotyledons of most seedlings emerge above the surface of the soil and acquire a green color.

Physiology of flowering, pod formation and flower drop factors.

Flower drop begins at the lower nodes and continues upwards. It has been observed that normal flower maturation per inflorescence averages 2.8 days and the gradation between the flowering of two neighboring internodes averages 1.9 days. However, different factors can influence this pattern: maximum temperatures on the day of flowering, or the day before, as well as the average temperature are closely related to the number of flowers. The reduction of temperature from 21°C during the day and 16°C at night to 14°C and 9°C respectively for 10 days considerably reduced the fall of young pods and flowers (Camarena, 1981).The absence of calcium affects the meristematic regions of the stem, leaves and root, which easily die early, mitosis stops and young leaves show poor formations. In the end, leaves fall and apex growth stops; it can also affect other young organs in formation such as fruits, which usually present apex degeneration and a lower resistance to fungal infection as cited in (Martinez, 1995) by **Camarena et al, (2003).**

Boron physiology in plants.

The physiological functions of boron are not yet fully elucidated. Its role in plant metabolism is perhaps the most unknown of all the essential nutrients, despite being the micronutrient with the highest molar concentrations, at least in dicotyledons. Boron always acts with valence III, so it does not intervene in any redox process inside plants. It has not been found to form part of any enzyme system, although it acts as a modulator of enzymatic activities. It has also been shown that, in certain cases, it can be partially substituted by germanium, aluminum or silicon. All of the above does not mean that it does not perform essential biological functions for the plant. As we will see below, boron plays an essential role in sugar transport, sucrose synthesis, protein metabolism, synthesis and stability of cell walls and membranes, etc., quoted from (VERA, 2001) by **Camarena et al, (2003).**

2.3.- Conceptual Bases

2.3.1. Different concepts and definitions of irrigation by solarization

NOVAGRID (2014) refers in its article published on Friday, November 14, 2014 that irrigation by solarization is an irrigation technique used in **small orchards,** where the upper part of 5 liter PET bottles normally used for bottling water is used as a plastic cover, inside which is placed the base of a smaller PET bottle (1.5 liter) that acts as a container for irrigation water. The main advantage of this system is that it is an inexpensive system when used in **small areas**, it is easy to handle and only requires refilling when the water container is empty. It also allows the water bottles to be reused.

SITIO SOLAR.COM (S.F.) details that, the solar drip, also known as **Kondenskompressor,** is an irrigation technique that allows to achieve an optimal use of water by using the sun's energy as a driving force in the process of distillation and movement of water. It is a system of surprising simplicity and efficiency by means of which it is possible to reduce the amount of irrigation water by up to 10 times compared to traditional irrigation systems.

SANTA CRUZ (2017) notes that, drip irrigation allows a constant supply of water to the plants during a certain period of time, which makes it possible not to worry about returning from vacation and finding the plants dry or withered. In addition, **with the realization of these irrigation systems also contributes to the care of the environment,** through the recycling of the materials used, as well as to the saving of water since, in this way, the plant only consumes the amount that it needs and, at the same time, it avoids the excessive expenditure of this so precious good. The construction process is very simple. We start with a plastic bottle of a liter and a half or two liters, depending on the time we want to have the system in operation. Then, the lid is removed and one or more holes are drilled with a nail or a drill. The number of holes drilled in the lid will determine the rate at which the water flows, so the more holes, the faster the water will drip. The size of the hole will also contribute to the rate of water flow: a smaller hole will allow the water to drip slowly, while larger holes will cause the water to run out quickly. The next step is to make a hole near the plant or group of plants, which should be deep enough to bury the mouth of the bottle and keep the bottle filled with water in an upright position. All that remains is to secure the bottle in place by pressing the soil around it and let the system and gravity do their work. These irrigation systems can also be placed aerially, by simply placing the bottle above the plant to be maintained with a constant flow of water.

ZACARIAS (2020) describes that the solar drip, also known as **Kondenskompressor**, is an irrigation technique designed to achieve an optimal use of water by using the sun's energy as a driving force in the process of water distillation and movement. It is a system of surprising simplicity and efficiency by means of which it is possible to reduce the amount of irrigation water by up to 10 times compared to traditional irrigation systems. The solar drip irrigation system with bottles also has the advantage of making it possible to use brackish water or even seawater for irrigation, since this process transforms salt water into fresh water. In the manufacture of the **kondenskompressor** a very abundant and easy to obtain material such as PET plastic bottles can be used. Its manufacture and installation is very simple and is within the reach of any farmer, whether in a domestic or professional environment. It also requires very little maintenance, being only necessary to replenish the water tank when necessary and remove the plants that may have grown inside the **Kondenskompressor.** With the application of this technique, the plants develop fully, using only the necessary amount of water and avoiding the evaporation of water that is not used. Because it requires materials that are very abundant waste and because both manufacture and installation are extremely simple, this technique can be very easily employed in poor countries with prolonged dry seasons. Even drip irrigation with bottles can be used in desert areas with access to some source of fresh or salt water (e.g., near the sea).

2.3.2.- Operation

ZACARIAS (2020) narrates that the solar drip technique works in a very similar way to solar stills, using the sun's energy to evaporate the water in an area and by means of the shape of the device to direct it to where it is of interest. When the sun's rays hit the **Kondenskompressor,** the greenhouse effect is produced inside it, raising the temperature of the air and causing the water in the tank to evaporate. The air inside the hood becomes saturated with humidity, causing condensation in the form of droplets. on the wall. As long as the **kondenskompressor** is exposed to the sun, evaporation continues and larger and larger droplets are formed, which eventually slide down the walls and fall on the ground, watering it. In this way the natural water cycle is reproduced on a small scale.

2.4. Definition of Terms

▶ **Database: An** organized and integrated set of data stored in a computer, in order to facilitate its use for multipurpose applications.

▶ **Hydrological cycle:** In this process, the protagonist is the water as it undergoes various transformations that occur due to some physical and chemical reactions. This whole process also includes displacements that make water pass through various states of matter such as solid, liquid and gaseous.

▶ **Water consumption:** In water statistics and water accounts, the concept of **"water consumption"** is used as an indicator of water consumption. consumption indicates the amount that is lost to the economy during its use, in the sense that it enters the economy but does not return to water resources or the sea; this happens because during water use, some of it is incorporated into products; it evaporates; it is transpired by plants; or it is simply consumed by households or livestock.

▶ **Condensation:** Condensation is when water arrives in vapor form at. The condensation phase occurs at high altitudes, when it returns to its liquid form and falls.

▶ **Distillation: A** process that consists of heating a liquid until its The main objective of distillation is to separate a mixture of several components by taking advantage of their different volatilities, or to separate the volatile materials from the non-volatile ones. The main objective of distillation is to separate a mixture of several components by taking advantage of their different volatilities, or to separate the volatile materials from the non-volatile ones.

► **Efficiency:** Ability to dispose of someone or something to achieve a given effect.

► **Evaporation:** Evaporation, gradual conversion of a liquid into a gas without boiling. The molecules of any liquid are in constant motion. The average (or mean) velocity of the molecules depends only on temperature, but there may be individual molecules moving at a much faster or much slower rate than the average. At temperatures below the boiling point, it is possible that individual molecules approaching the surface with a higher than average velocity may have enough energy to escape from the surface and pass into the space above as gas molecules.

► **Scale:** The physical dimension, either in space or time, of the observation of phenomena.

► **Solar drip**: Solar drip is the name given to the system that traps evaporated water from the ground and from the reservoir inside a KondensKompressor by the action of solar energy. The moisture condenses on the inside walls and falls in the form of solar drip directed through the walls back to the ground.

► **Kondenskompressor:** It is a technique that produces distilled water with solar radiation. It is a very simple and effective system that produces a "solar drip" by means of which irrigation water can be reduced 10 times compared to traditional systems. We only need to reuse a few plastic bottles.

► **Model:** Simplified representation of a limited part of reality and related elements.

► **PET bottle:** PET (polyethylene terephthalate) was first produced in 1941 by British scientists Whinfield and Dickson, who patented it as a polymer for the manufacture of fibers. PET is used to manufacture bottles because of its high resistance to chemical agents, great transparency, light weight and lower manufacturing costs. PET plastic is the The most recycled plastic in the world, precisely because it is not recommended for reuse and is very resistant to biodegradation. If this plastic is abandoned in nature, it can take between 100 and 1,000 years to decompose, so it is essential to recycle it.

► **Growing Period (GP):** Period of the year in which the growing conditions of a plant are (See Table 2 for definition and types of growing seasons and their components).

► **Potential agronomic yield: The** maximum yield that can be obtained from a plant. achieved by a given crop in a specific area, taking into account the biophysical constraints, preferably of climate and soil.

▶ **Potential yield:** Maximum yield that can be achieved by a variety of a given crop in a specific area, as a function of radiation and temperature.

▶ **Phenological requirement:** Requirement of a crop in terms of environmental conditions necessary for its development, considered within the development cycle of the crop.

▶ **Irrigation: The** supply of water to the soil by various methods to facilitate the development of plants. It is practiced in all parts of the world where rainfall does not supply sufficient moisture to the soil or where irrigated crops are to be planted. In dry areas, irrigation should be used as soon as the crop is planted. In regions with irregular rainfall, it is used during dry periods to secure crops and increase yields.

▶ **Management system: A** system composed of soil, crop, weed, pest and disease aspects, capable of transforming solar energy, water, nutrients, tillage and other inputs into food, feed, fuel or fiber. The management system is equivalent to a subsystem of the farming system.

▶ **Solarization: The** process by which the sun's rays are incident on a surface by increasing the temperature.

▶ **Transpiration:** It is the water that penetrating through the roots of plants is used in the construction of tissues or emitted by the leaves and reintegrated into the atmosphere. Transpiration depends on the type of plant, the evaporation power of the atmosphere, the degree of soil humidity, etc.

▶ **Sustainable land use:** Use of land that does not progressively degrade its

2.5.- Formulation of the hypothesis

Hypothesis

▶ Ho = The solar irrigation system is not efficient in the cultivation of fava beans in the province of Acobamba - Huancavelica.

Alternating hypothesis

▶ Ha = The solar irrigation system is efficient in the cultivation of fava beans in the province of Acobamba - Huancavelica.

2.6.- Variables

2.6.1. Variables independent

► Depth of cover levels in the solarization irrigation system

2.6.2. Dependent variable

✓ Quantity of water consumed

► Performance

2.6.3. Variable Intervening

► Weather

Operationalization of variables

Variable	Operational Definition	Indicator	Category or Scale	Measurement criteria categories
a. Independent Variable				
Depth of PET bottles.	Longitudinal measurement	m.	0.06 - 0.12 - 0.18	Experimental field
b. Dependent Variable				
Water consumed	Direct measurement	Liters/ treatments	Average	Crop cycle
Performance	Harvest	t / ha	average	Crop harvesting

Source: Own elaboration

CHAPTER III

RESEARCH METHODOLOGY

3.1. Temporal scope and spatial scope

3.1.1. Political location:

Region: Huancavelica.

Province: Acobamba.

District: Acobamba.

Location: "Vista Hermosa Casa Blanca".

3.1.2. Geographic location:

Altitude: 3423 meters above sea level.

South Latitude: 12° 50' 37.32" from the equator

Longitude west: 74° 34'41.46" Greenwich Meridian

3.1.3. Climatic factors:

Average annual rainfall: 650 ml

Relative humidity 60 %

Average annual temperature: 12°C

3.2. Type of Research

The research was **exploratory** because it corresponded to a pilot study that sought to achieve a first scientific approach to the study of the water needs of the bean crop conducted with solar irrigation using PET bottles in the environmental scenario of Acobamba large campaign December - June.

3.3. Level of Research

The present exploration work determined an **explanatory** level of causal relationship since it sought to evaluate the yield of the bean crop by relating its response to irrigation by solarization at three levels of depth of the PET bottles.

3.4. Method of Research

The scientific research method was based on work and evaluation of irrigation by solarization of fava beans using PET bottles installed at different depths of cover and, in the cabinet, evaluating, tabulating and inferring the results.

3.5. Population, Sample and Sampling

a. Population:

The data on water consumption and yield of the bean crop selected for the evaluation included information obtained by measuring with PET bottles and weighing the green beans obtained.

b. Sample:

According to **Oseda (2008:122) he** mentions that "the sample is a small part of the population or a subset of it, which nevertheless possesses the main characteristics of the population. This is the main property of the sample (possessing the main characteristics of the population) that makes it possible for the researcher, who works with the sample, to generalize his results to the population".

c. Sample size:

The sample size under study was:

STRATUM	Sample size
PET water tank	
PET cover	
Number of plants / stroke	
Total number of plants	

d.- Sampling:

Probability or stratified random sampling was used, because all the data on solarization irrigation in broad bean had the same probability of being chosen to form the sample.

3.6. Techniques and instruments for data collection

For the collection of primary data, the main unit of information taken was the volume of water evaporated from the water tank inside the PET casing of each treatment, which was obtained by means of a graduated jar.

3.7. Processing techniques and analysis of data

Excel (spreadsheet) was used to process the information, which allowed us to review and verify the data obtained with the instruments used in this research, as well as to calculate the water consumption for the bean crop based on the data obtained with the instruments used in this research. For the analysis and interpretation of the data we used summary data measures and to verify our hypothesis we made use of measures of association taking into account the nature of the study variables. (Statistical test not for Chi-Square metric with an $\alpha = 0.05$).

CHAPTER IV

PRESENTATION OF RESULTS

4.1. Analysis of the Information 4.1.1.1.

In the present study, considering the history of the land whose previous planting was potato, fertilization was by incorporating decomposed organic matter at a rate of 50 g. per plant, which represents 1.3 t./ha., considering a density of 27778 (0.9 m., between rows and 0.4 m., between plants), with three plants/ha.

4.1.2.- Emergence and initial development of the broad bean crop.

The sowing of the broad bean variety marbled was carried out on January 14, 2020, depositing 04 seeds per hole, which corresponds to a density of 27778. The emergence of the seedlings once the seeds germinated began on January 26, 2020 (12 days after sowing), it should be clarified that the seed had a pre-germination for which it was soaked in water for 48 hours before sowing. Bean seedlings were thinned on February 12 when they were approximately 10 cm high, leaving the three most vigorous plants in each hole.

4.1.3. Height of broad bean plants

The sowing was carried out on January 14, 2020, and the development of three bean stalks was manifested after thinning:

Table No. 01. Average plant height of broad bean in days

Date	m	days
24/03/2020	0,65	
13/04/2020	1,05	90
28/04/2020	1,10	
21/05/2020	1,45	
01/06/2020	1,68	

Own elaboration

As can be seen in Table No. 01, the maximum height was obtained 130 days after planting represented 1.68 m., on average, this height is higher than the characteristic of the variety Pacae jaspeado obtained in plantings under irrigation in the Mantaro Junín valley, probably this is due to the water available in the root zone as a result of the irrigation system by solarization that was implemented in this study that offered water in rational amount to the bean plants.

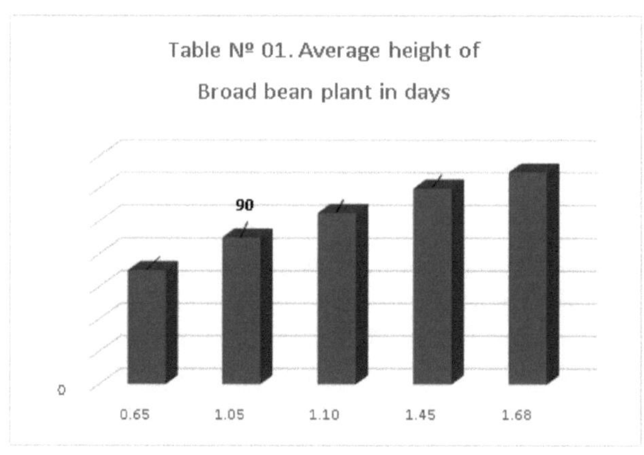

Own elaboration

These results are objectively shown in Figure No. 01, which achieves the crop in the period of flowering - maturity of the crop.

Table No. 02.- Average number of fava bean tillers/strike

Date	N°.	days
13/04/2020		90
28/04/2020		

Own elaboration

As a result of evaluating the cultivation process in relation to the number of tillers obtained, the average number of tillers obtained was 12 at 105 days after planting, which gave us stems and tillers with a firm structure located in the neck of the plant, with an average of 4 tillers per plant.

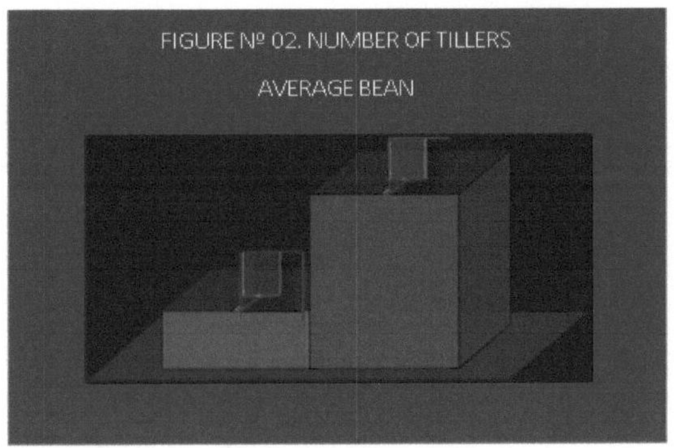

Own elaboration

4.1.4. Number of Pods per Bean Plant

The beginning of flowering was on March 22 culminating on May 05, 2020, it should be noted that 100% soil coverage occurred 107 days after planting (30 / 04 / 2020), the total number of pods evaluated on average per stroke was 85 of which manifested its productive viability 20% (17 mature pods).

4.1.5. Bean Plant Yield per Treatment

The average yield obtained per hit of three green bean seeds was **0.34 kg** (20 gr., average weight per harvested pod), although there was a 2% loss due to bird damage.

4.1.6 Biological harvesting of fava beans using a solar irrigation system.

To determine the biological harvest of the bean crop, roots, stems and leaves were taken from the experimental field in a discriminate manner for each hole to determine their dry matter in the laboratory.

4.1.7 Water demand for solar irrigation system in broad bean.

The solarization irrigation system consisted of two parts: a water reservoir and an external covered lid that was placed near the bean plant to supply moisture within the radius of its roots. To operate the system, the reservoir (a 1-liter disposable cup) was filled with water that, when heated by the sun's rays through the The evaporation of water from the glass and covered soil condensed on the inner lateral and upper wall of the cover, forming small drops of water that then by the action of gravity slid down the cover of the PET bottle, returning to the ground, thus producing the natural water cycle in a controlled manner throughout the vegetative period of the bean crop. In order to avoid the influence of rain precipitation, a temporary cover was installed at each rainy season.

Table No. 03 Comparative table of water consumption/crop cycle

Solar irrigation system			
Period	0,06	0,12	0,18
First			
		320	
Second		320	
	550		225
Total, ml:	425	257	
Total l.:	0,425	0,257	0,175
Total, m3:	0,000425	0,000256667	0,000175
Total, m3/ha:	11,81	7,13	4,86

Source: Own elaboration

(a). - Solar irrigation 0.06 depth (b). - Solar irrigation 0.12 depth (c). - Solar irrigation 0.18 of depth From Table N° 03 it can be deduced that the highest water consumption using the solarization irrigation system corresponds to the bean crop where the solar meter was installed at 0.06 m above ground level, with a demand during the entire crop cycle of 0.425 liters/planting (three bean plants).

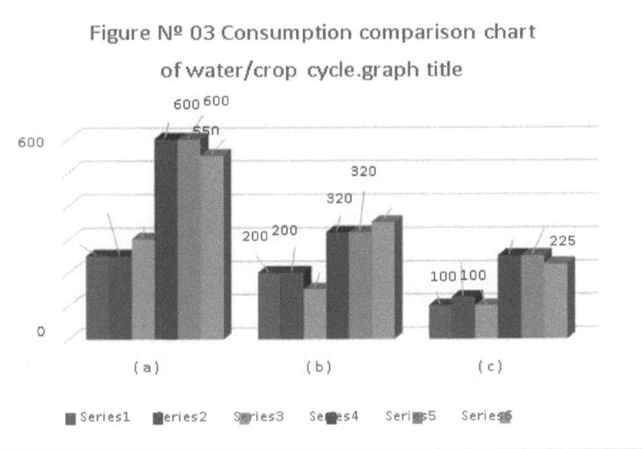

Figure N° 03 shows that the highest water demand by the crop occurs from the middle cycle to the harvest of the crop.

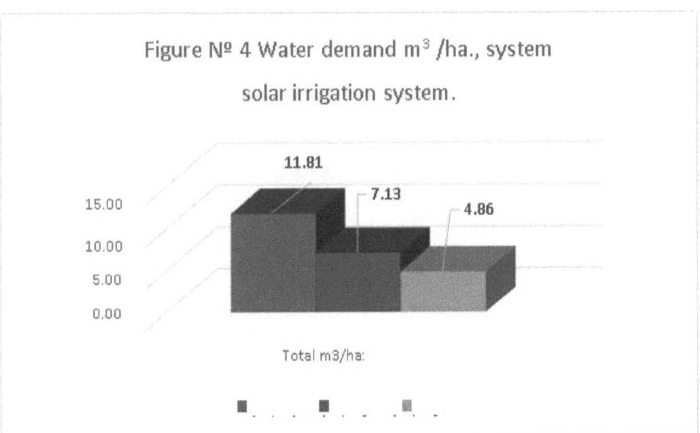

Figure N° 04 shows that the water demand through the solarization irrigation system in the bean crop located at 0.06 m. from the ground level per hectare is only 11.81 m3/ha. during the whole crop cycle.

4.2. Hypothesis testing. -

Table N° 01 Hypothesis test of the thesis "Efficiency of the irrigation system by solarization at different depths in the cultivation of broad bean (Vicia faba L.) in the province of Acobamba".

$$CT > T \text{ (Table)}$$

Chi-square test of goodness of fit

$$TC = 2.12 > -0.8076 \text{ y} - 1.2666$$

Ho is rejected and Ha is accepted.

Hypothesis:	Ha = The solar irrigation system is efficient in the cultivation of fava beans in the province of Acobamba - Huancavelica.
The test statistic value is greater than the tabular value, the null hypothesis (HO) is rejected.	

Source: own elaboration

For this analysis we used Students' t-distribution table, where for the degree of freedom found = 2.12 we had to interpolate the immediate upper and lower one, obtaining values for the following two-tailed probability 0.05 = - 0.876 and 0.01= - 1.2666.

4.3. Discussion of results

4.3.1.- Emergence and initial development of the broad bean crop.

It was demonstrated that the germination and uniformity of bean plant emergence in the field are favored by a pre-germination of the seeds for 48 hours prior to sowing, which, together with irrigation by solarization, favors this process.

4.3.2. Bean plant height

As described in the corresponding item after thinning, the three best plants that had germinated per sowing stroke were left, it was possible to observe a subsequent development of stems and birth of tillers (105 days after sowing) with firm and solid structure located in the neck of each plant with an average of 4 tillers, the bean crop reached its maximum height at 130 days of its vegetative period represented in 1.68 m., it is inferred that the average plant height of bean obtained through the system The results of the solar irrigation system are superior to those of the Pacae jaspeado variety planted in the same season under irrigation in the Mantaro Junín valley, which shows that the solar irrigation system provides moisture "water" to the soil in a rational, permanent and timely amount to the crop.

4.3.3. Number of Pods per Bean Plant

The solarization irrigation system allowed a productive viability of 20% of the bean crop, which favors the formation and filling of pods in a uniform manner, coinciding this stage with the total soil cover, this situation allows us to state that this irrigation system ensures the regular development of the crop during its different stages without detriment to production.

4.3.4. Bean plant yield per treatment

It can be deduced that the production of green beans on average using three seeds per stroke ensures a yield of **9.25 t/ha, even with a 2% loss due to bird damage.** Even with a 2% loss due to bird damage, this result of the solarization irrigation system offers advantages to the producer who cultivates under rainfed or rainfed conditions.

4.3.5 Organic harvesting of fava beans using solar irrigation system

The biological harvest of roots, stems, leaves and fruits of the broad bean crop represented an average of 1,600 kg/stalk (three seeds) after the broad bean harvest, which expresses a normal productive development of the crop.

4.3.6.- Water demand in bean solar irrigation system.

The water demand through the solarization irrigation system in the bean crop located with cover at 0.06 m above ground level per hectare is 11.81 m3 /ha, at 0.12 m, it is 7.13 m3/ha, and 4.86 m3/ha, at 0.18 m, during the entire crop cycle, which means that the average consumption was **7.93 m3/ha.**

4.3.7.- Hypothesis test of the thesis "Efficiency of the irrigation system by solarization at different depths in the cultivation of fava bean (Vicia faba L.) in the province of Acobamba". Analyzed the hypothesis test of the present thesis, it is stated that the established methodology (proposal) to determine the efficiency of the irrigation system by solarization carried out at different depths in the bean crop shows that it is very efficient, simple and economical, besides being very easy to install, recycling PET bottles, It is an optimal irrigation technology for the production of crops under the rainfed production regime, increasing production and reducing the demand for irrigation water to a minimum level of requirement per hectare with 95% and 99% of probability, so it is inferred that it is a practical irrigation method for the cultivation of areas where water is scarce.

CONCLUSIONS

1. It is inferred that the solar irrigation system allows efficient irrigation of the crop, being its installation in the field very simple and without any economic cost, since recycled materials from **PET** bottles are used.

2. It is concluded that irrigation by solarization implies a water saving greater than that obtained with the use of

100% with respect to the exclusive use of traditional irrigation systems.

3. The water demand through the solarization irrigation system in the broad bean crop varies from 0.06 m., from the soil level per hectare is 11.81 m3/ha. at 0.06 m., 7.13 m3/ha. at 0.12 m., and 4.86 m3/ha. at 0.18 m., during the entire crop cycle, which means that the average consumption was **7.93 m3/ha.**

4. From the hypothesis test carried out, it is theorized that the methodology established (proposed) for irrigation by solarization in the bean crop is very efficient, simple and economical, as well as very easy to install, recycling PET bottles of 5 liters capacity, characterizing an optimal irrigation technology for the production of crops under the rainfed production regime, **favoring an increase in production and decreasing the demand for irrigation water** to a minimum level of requirement per hectare with 95% and 99% probability.

5. We can conclude by inferring that the main advantage of this solar irrigation system is that it is an inexpensive system when used in small areas, it is easy to operate and only requires refilling the evaporation cup with water when it decreases.

RECOMMENDATIONS

1. It is urged to carry out similar work to **study the solar irrigation system** for other crops in the Province and District of Acobamba Huancavelica.

2. In order to avoid errors in the calculation of water demand for irrigation using the solarization system, it is recommended to use PET bottles of homogeneous capacity to guarantee favorable results.

3. It is recommended to promote the reuse of water bottles made of petroleum and carbon called PET, which do not dissolve, since recycling this material saves energy and natural resources, taking into consideration that plastic is a serious threat to the environment because it takes years to degrade.

4. As a final consideration, I consider that the solar irrigation system in other crops in the Province and District of Acobamba Huancavelica should be a priority topic to be investigated by graduates of the EP Agronomy.

REFERENCES

1. **CAMARENA MAYTA, Félix; CHIAPPE VARGAS, Luis; HUARINGA Joaquín, MOSTACERO NEYRA, Elvia (2003).** Broad bean cultivation manual. La Molina / PERU.

2. **FAO, FAO Irrigation and Drainage Study - Guide 56 (1990),** Guide for the determination of crop water requirements, 322 pp.

3. **FLORES RIOS, Arnoldo Walter (2016), Thesis**: Universidad Mayor de San Andrés Facultad de Agronomia carrera de Ingenieria Agronomica "Estudio del uso de botellas plasticas recicladas **(Politereftalato de etileno)** (PET) en el riego por goteo solar y su aplicaciòn en la forestaciòn", Bolivia, 119 pp.

4. **GARAY CANALES, Oscar Baldomero (2009),** Manual de uso consuntivo del agua para los principales cultivos de los Andes Centrales Peruanos; INCAGRO, 34 pp.

5. **MARMOLEJO GUTARRA, Doris and SUASNABAR ASTETE, Carlos (2000).** Leguminosas de Grano. Ediciones "UNCP". Huancayo / PERU.

6. **ALIAGA BARRERA, Isaac N. (2004).** Botany Course. "Botanical Nomenclature". Booklet. U.N.H. - E.A.P.A. Huancavelica / PERU.

7. **MINAG - INSTITUTO NACIONAL DE RECURSOS NATURALES - IINRENA INTENDENCIA DE RECURSOS HIDRICOS OFFICINA DE PROYECTOS DE AFIANZAMIENTO HIDRICO, (2006);** Proyecto de Irrigación Molinos Volume II, 22 pp.

8. **JAIME PIÑAS, Jesús Antonio, BAUTISTA VARGAS, Marino and RUIZ VILCHEZ, David (2008),** Research Project "Evaluation of the water resources to be diverted to the Acobamba Irrigation System "49 pp.

9. **JAIME PIÑAS, Jesús Antonio, (2017),** "Demanda de Agua para Riego en la Sierra", Universidad Nacional de Huancavelica- Facultad de Ciencias Agrarias, 160 pp.

10. **JAIME PIÑAS, Jesús Antonio, (2014),** thesis **Validaciòn de propuesta Metodologica para el Diseño Hidrico de Proyectos de Riego en la Sierra. Peruana", Universidad Nacional de Huancavelica- Facultad de Ciencias Agrarias, 125 pp.**

11. **PADILLA, Veronica, CERON MATAMOROS, Itzae, HERNANDEZ LUNA, Mia Sarahi, GALVAN TELLEZ, Iktan, Muciza(2020),** Solar drip irrigation.

Web graphics

1. www.novagric.com/es/blog/articulos/sistema-de-riego-solar (2014) article irrigation.
2. http://www.sitiosolar.com/la-tecnica-de-riego-del-goteo-solar-kondenskompressor/.

3. https://www.santacruzlimpia.info/index.php/blog/item/362-riego-por-goteo-con- recycled-plastic-bottles.
4. **ZACARIAS PERDOMO** (2020), https://maestroviejo.es/sistema-de-riego-por- drip-solar-with-bottles/

CHI-SQUARE GOODNESS-OF-FIT TEST

Table N° 02 Comparative table of water consumption/crop cycle.

(a)	(b)	(c)
600	320	
	320	
550		225
2550,00	1540,00	1050,00
425,00	256,67	175,00
6,00	6,00	6,00
153750,00	34533,33	27500,00

Solar irrigation dc./treatment.			
Period	(a)	(b)	(c)
First			
Second	600	320	
	600	320	
	550		225

SUMAAverage =n1 =SC M1 =S^2 C =18828,33

Observation: Own elaboration.

T =	2,12
T 0,05 =	-0,8076
T 0,01 =	-1,2666

Remarks:

(a). - Solar irrigation 0.06 of depth (b). - Solar irrigation 0.12 of depth (c) - Solar irrigation 0.18 of depth (d) - Solar irrigation 0.18 of depth (e).

CT > T (Table)

Chi-square test of goodness of fit

TC =	2,12 > -0.8076 y - 1,2666
Ho is rejected and Ha is accepted.	
Hypothesis:	Ha = The solar irrigation system is efficient in the cultivation of fava beans in the province of Acobamba - Huancavelica.
The test statistic value is greater than the tabular value, the null hypothesis (HO) is rejected.	
Then it is concluded that the methodology established (proposed) for irrigation by solarization in the bean crop is very efficient, simple and economical, as well as very easy to install, recycling PET bottles, 5 l capacity, characterizing an optimal irrigation technology for crop production under the rainfed production regime leading to an increase in production and reducing the demand for irrigation water to a minimum level of requirement per hectare with 95% and 99% probability.	

Table N° 01.-CONSISTENCY MATRIX

PROBLEM	OBJECTIVES	HYPOTHESIS	VARIABLES	INDICATORS
General: How efficient will the irrigation system be	General: To know the efficiency of the	General: Ho:	Variable: Independent Depth levels of the solar irrigation system.	Measurement longitudinal (m). Direct measurement (liters/treatments).
By solarization a different	irrigation by solarization at different	The solar irrigation system	Amount of water consumed.	
depths in fava bean cultivation, in the	depths, in the cultivation of fava beans,	is not efficient in the cultivation of	Precipitation Temperature.	Crop cycle history (mm).
Province of Acobamba - Huancavelica?	in the province of Acobamba.	in the province of Acobamba.		Crop cycle history (°C).
		- Huancavelica.	Variable: Dependent Performance.	
Specific:	Specific:			
How could water loss be prevented	Evaluate water demand in the	Ha		
by evaporation in irrigation systems?	solar irrigation system in the	The solar irrigation system		
How would the amount of irrigation water be reduced in the cultivation of fava beans in the province of Acobamba? In what way could the waste be recycled?	province of Acobamba. Recycling of disposable beverage containers for use in irrigation in the	is efficient in the cultivation of fava beans, in the province of Acobamba - Huancavelica.		
disposable beverage containers in the	province of Acobamba.			
irrigation system, in the province of	Purify the irrigation water for the			
Acobamba?	crops in the province of Acobamba.			

50

TESTIMONY PHOTOGRAPHIC

Photograph N° 01

Water container for evaporation by solarization

Photograph N° 02

Pet covers for irrigation by solarization

Photograph N° 03

Bean plants in development

Photograph N° 04

Solar water meters located between plants.

Ensuring solar irrigation water solarization

Plants with solar solar irrigation water solarization

Printed by Books on Demand GmbH, Norderstedt / Germany